YOUR KNOWLEDGE HAS VALUE

- We will publish your bachelor's and master's thesis, essays and papers

- Your own eBook and book - sold worldwide in all relevant shops

- Earn money with each sale

Upload your text at www.GRIN.com and publish for free

Nino Crisolo

Discovering Ohm's Law. With Great Power Comes Great Current Squared Times Resistance

GRIN Verlag

YOUR KNOWLEDGE HAS VALUE

- We will publish your bachelor's and
  master's thesis, essays and papers

- Your own eBook and book -
  sold worldwide in all relevant shops

- Earn money with each sale

Upload your text at www.GRIN.com
and publish for free

Bibliographic information published by the German National Library:

The German National Library lists this publication in the National Bibliography;
detailed bibliographic data are available on the Internet at http://dnb.dnb.de .

Imprint:

Copyright © 2016 GRIN Verlag
Print and binding: Books on Demand GmbH, Norderstedt Germany
ISBN: 9783668824157

This book at GRIN:

https://www.grin.com/document/439060

Nino Crisolo

Discovering Ohm's Law. With Great Power Comes Great Current Squared Times Resistance

GRIN Verlag

Content

Overview

In exploring the world of electricity it is essential to start by understanding the basic concepts of current, resistance, and voltage or potential difference. These three key building blocks are required to manipulate and investigate electricity. Unseen concept like this can be detected by the use of measuring tools such as a ammeter, voltmeter and ohmmeter. This will help the students visualize what is happening with the charge in a system. The relationship between voltage, current and resistance will be explained thoroughly in this learning booklet.

1. Define Ohm's law;
2. Use Ohm's law to calculate current, voltage and resistance in simple electric circuits;
3. Determine the current, voltage and resistance using measuring devices;
4. Calculate the power of a circuit given any two of the three electrical quantities – current, voltage and resistance.

Ohm's Law

In 1827 a German physicist, Georg Simon Ohm experimentally established the relation among electric current, resistance and potential difference in an electric circuit. He found out that the current passing through conductor varies directly as the potential difference applied at its ends and inversely as the resistance of the conductor. This statement is called Ohm's law and may be applied to the whole circuit or to a particular part of a circuit.

Applied to the whole circuit $I_T = V/R_T$, where I_T is the total current, V is the voltage, and R_T is the total resistance of the circuit.

Applied to a portion of the circuit, $I = V/R$ where I is the current in ampere (A) passing through that part of the circuit, V is the potential difference in volts (V), and R is the resistance in ohms (Ω) of the same part of the circuit.

Cover the variable you want to find and perform the resulting calculation *(Multiplication/Division)* as indicated.

While we can calculate the current, potential difference and resistance mathematically, there are devices which give us measurements of the these three quantities in a circuit. The three devices are:

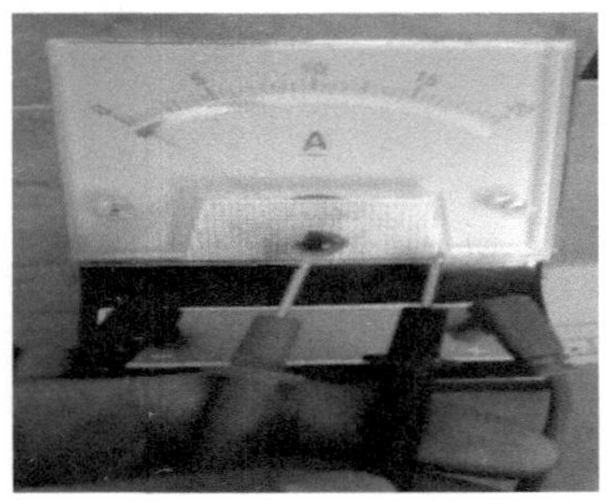

Ammeter- device used to measure current in amperes (A) (or milliamperes).

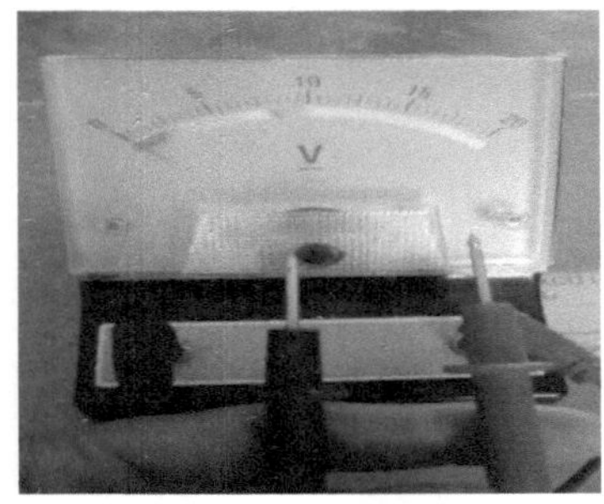

Voltmeter- device used to measure voltage in volts (V).

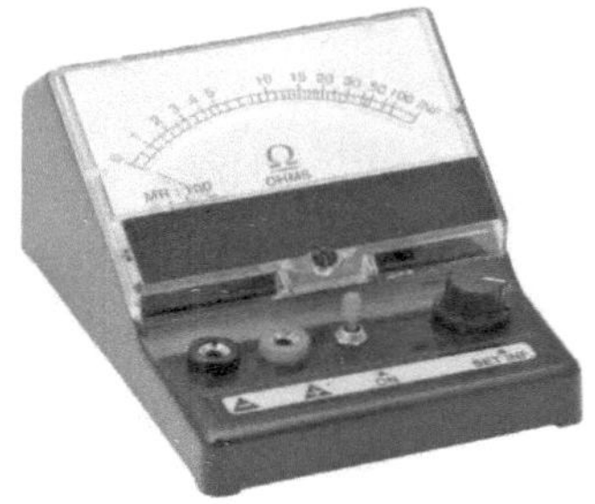

Ohmmeter- device used to measure resistance in ohms (Ω).

Sample Problem 2.0

1. How much current flows through a lamp with resistance 85Ω when it is connected to a 220-V outlet?

> *Given:* $R = 85\Omega$
> $V = 220V$
> *Find:* I
> *Solution:*
> $$I = \frac{V}{R}$$
> $$= \frac{220 \text{ V}}{85 \text{ }\Omega}$$
> $$I = 2.59 \text{ A}$$

2. What is the resistance of a lamp which allows 0.7 A current when 110.0 V is applied to it?

> *Given:* $I = 0.7 \text{ A}$
> $V = 110 \text{ V}$
> *Find:* R
>
> *Solution:* :
> $$R = \frac{V}{I}$$
> $$= \frac{110 \text{ V}}{0.7 \text{ A}}$$
> $$R = 157.14 \text{ }\Omega$$

3. While repairing an electric bulb socket, an electrician gets a mild shock when a current of 0.004 A passes through him. The same electrician is killed by a current of 0.16 A when he turned on an electric bulb while taking a bath. The voltage in each situation is equal to 120 V. Find the resistance of the electrician in each situation.

For the first case	*For the second case*

Given: $I = 0.004 \text{ A}$
           $V = 120 \text{ V}$
 Find: R
 	
 Solution: :
 $R = \dfrac{V}{I}$
 $= \dfrac{120 \text{ V}}{0.004 \text{ A}}$
 $R = 30{,}000 \text{ }\Omega$
	Given: $I = 0.16 \text{ A}$
	           $V = 120 \text{ V}$
	 Find: R
	
	 Solution: :
	 $R = \dfrac{V}{I}$
	 $= \dfrac{120 \text{ V}}{0.16 \text{ A}}$
	 $R = 750 \text{ }\Omega$

Note: The human body's resistance to current is in the order of 5000 000 Ω when the skin is dry. This resistance decreases when the skin is wet which can go as low as 100Ω when it soaked with saltwater. This is because ions in saltwater are current carriers and readily conduct electric charge.

The lie detector measures several parameters. One of which is skin conductivity or skin resistance. This based on the premise that a person sweats more when under stress and thus affects his or her resistance to current.

Activity 2: With great power comes great current squared times resistance

Key Learning Points

1. Learn about Ohm's Law.
2. Be able to use a two different device to collect data.
3. Explore the concepts of voltage and current.

Introduction

In this activity, students will be able to manipulate the data by simply changing the value of the resistor in the circuit. Therefore they are expected to figure out the relationship between voltage, current and resistance which is clearly stated in Ohm's Law.

You will need:

1. Ammeter (range of 1A)
2. Voltmeter (range of 10-15V)
3. 2 20Ω resistors
4. 2 30Ω resistors
5. Alligator clip
6. 4 1.5 dry cells

Experimental setup:

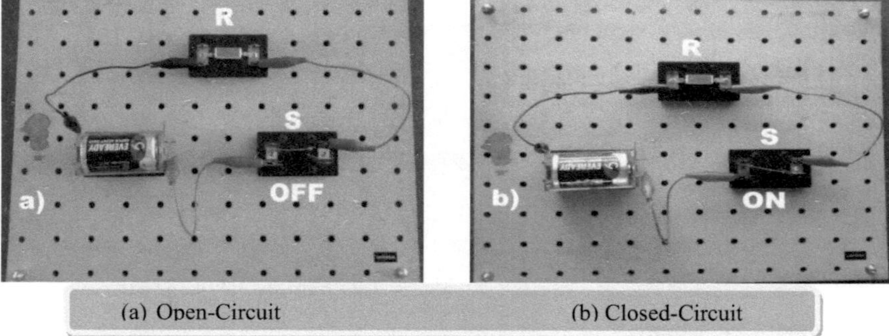

(a) Open-Circuit   (b) Closed-Circuit

Procedure:

1. Set up the circuit as shown above. For the first set of data, keep the resistance constant, using only one 20Ω resistor. Using one dry cell in the circuit, get the reading of the ammeter and voltmeter. Repeat, adding one dry cell at a time until all the four have been connected. Get the reading after adding each dry cell. Record the data in Table 1.
2. With the four dry cells in the circuit as the constant factor vary the number of resistors to be connected. For the first trial use one 20Ω resistor. Get the reading of the ammeter and voltmeter. For the second trial, use two 20Ω resistors; for the third trial two 30Ω resistors; and for the fourth trial two 30Ω and one 20Ω resistors. Record the reading for every variation of the number of resistors in table 2.
3. Make graphical representations of each table to visualize the relationship between the given variables.

Table 1 Comparative Values of Voltage and Current at Fixed Resistance

Electrical Quantities	Circuit Reading			
	1	2	3	4
Current (A)				
Voltage (V)				
Resistance (Ω)				

Table 2 Comparative Values of Current and Resistance at Fixed Voltage

Electrical Quantities	Circuit Reading			
	1	2	3	4
Current (A)				
Voltage (V)				
Resistance (Ω)				

Observations:

1. From table 1, how does current (I) change as the voltage (V) increases?
2. From table 2, how does current (I) change when resistance (Ω) increases?

Results and Analysis

In order to measure small amount of current a digital multi tester was used instead of an analogue ammeter.   It also helps the reader in   getting   the data more accurately . The result was also manually   computed and verified using a java application from phet.colorado.edu. This also includes a graphical representations of each table to analyze and interpret the data much easier. The outcome of this experiment gives evidence on the relationship between voltage, current and resistance.

First set up (Circuit diagram)

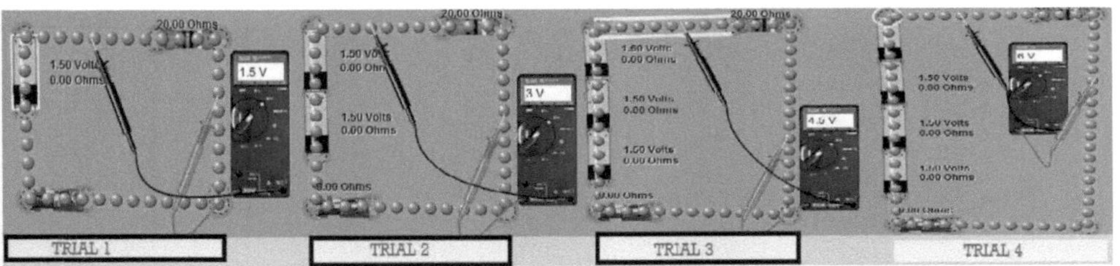

Table 1 Comparative Values of Voltage and Current at Fixed Resistance

Electrical Quantities	Circuit Reading			
	1	2	3	4
Current (A)	0.07	0.16	0.22	0.30
Voltage (V)	1.5	3	4.5	6
Resistance (Ω)	20	20	20	20

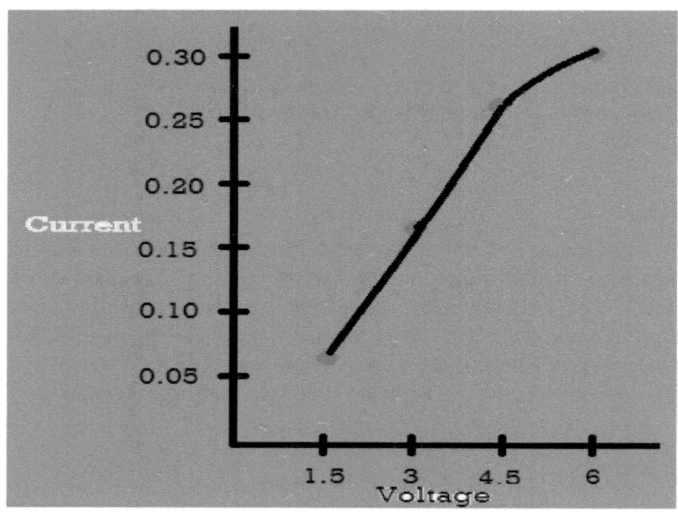

Figure 1-
Current versus Voltage
at constant Resistance

Second set up (Circuit diagram)

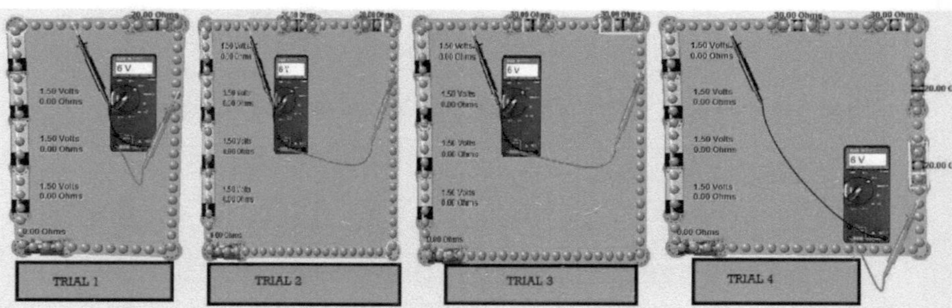

Table 2 Comparative Values of Current and Resistance at Fixed Voltage

Electrical Quantities	Circuit Reading			
	1	2	3	4
Current (A)	0.30	0.15	0.10	0.06
Voltage (V)	6	6	6	6
Resistance (Ω)	20	40	60	100

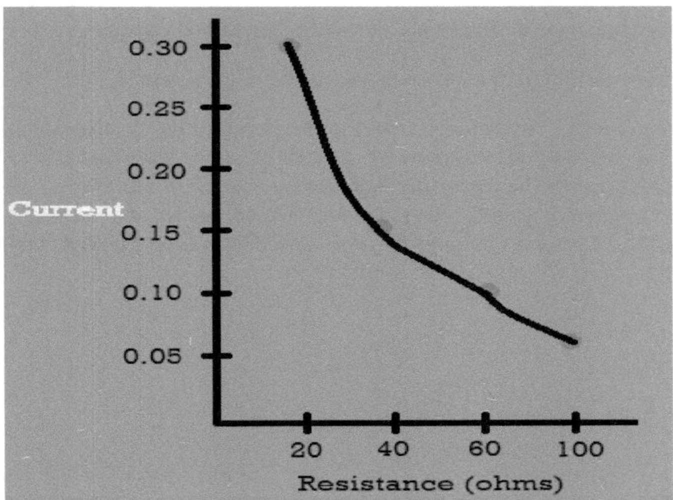

Figure 2-
Current versus Resistance
at constant Voltage

Observations:

1. From table 1, how does current (I) change as the voltage (V) increases?

   "The results shows that putting additional battery on the circuit meaning inducing the voltage, makes the current increase. Circuit diagram on the first set up illustrate that doubling of the battery voltage leads to a doubling of the current in the circuit. Figure 1 also visualize that current is directly proportional to voltage at constant resistance. The greater the voltage the greater the current. This proven the ohm's law equation I=V/R which represent the direct proportionality between current and voltage."

2. From table 2, how does current (I) change when resistance (Ω) increases?

   "Data shows that current decreases when the resistance is being increase at constant voltage. The increase in resistance by a factor of two as illustrated on the second circuit diagram causes the current to decrease by a factor of two to one-half of its original value. Figure 2 clearly represent the inverse relationship between current and resistance, as for the one quantity (resistance) increases the other one (current) decreases."

Application

Ohm's law helps us in determining either voltage, current or resistance of a linear circuit when the other two quantities are known to us. Apart from that, it makes power (rate of energy transfer) calculation a lot simpler, like when we know the value of the resistance for a particular circuit, we need not know both the current and the voltage to calculate the power dissipation since P = VI. Rather we can use Ohm's Law.

$$V = I_R$$

$$I = \frac{V}{R}$$

To replace either the voltage or current in the above expression to produce the result

$$Thus,\ P = VI = \frac{V^2}{R} = I^2 R \quad (Since\ V = IR)$$

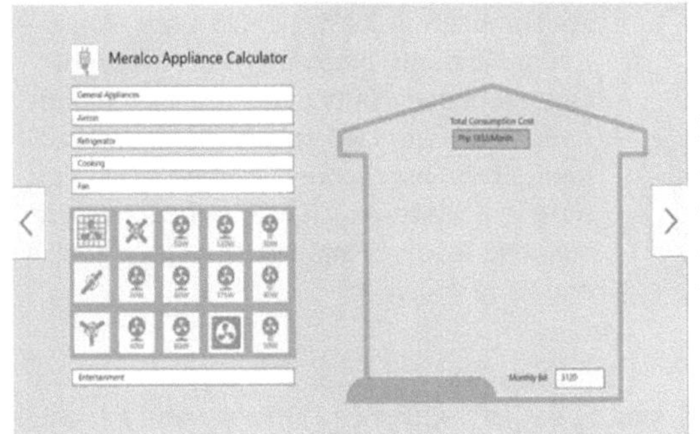

Meralco Appliance Calculator is an application used to calculate the operating cost of our home appliances and tracking which among our home appliances contribute most to our monthly electric bill.

This application is very helpful in monitoring our electric charges, however do you know that using Ohm's law we can also compute and determine the cost of our monthly bill? Here is how, since we already know that power is the product of current and voltage. Therefore energy can also be expressed using current, voltage and time. In equation form this is given as W = PΔt. Now the unit of energy used by these companies to calculate our consumption is the kilowatt-hour (kWh). This is the energy delivered in one hour at the constant rate of 1 kW.

Let us try to calculate the cost to operate a 20" desk fan for 12 hours, it has a wattage value of 78 W and the   electric energy cost is 4.0004 /kWh pesos (based on 02-03-2016 Meralco bill).

> *Given:* cost of energy = P 4.0004/kWh
> P= 70W or 0.079kW
> Δt=12h

> *Find:* cost to operate the fan for 12 hrs
> *Solution:* W = PΔt
> = (0.079kW) (12h)
> W = 0.948kWh
>
> *Cost = (0.948 kWh) (P 4.0004/ kWh)*
> **Cost = P 3.79**

Electric meter measures our electric consumption in kWh, or Kilo Watt Hours. Every time an electric meter measures One thousand Watts of electric use in one hour, the meter will record one kilo watt hour. This may sound complicated but after calculating a few products, it will become easy. Remember 1000 watts equals to 1 kilowatt.

Calculating electric usage is just one way of applying Ohm's law. In fact manufacturer of certain appliances consider this three variables voltage, current and resistance in creating useful machine. For example by increasing or decreasing the amount of resistance in a particular branch of the circuit, a manufacturer can increase or decrease the amount of current in that branch. Kitchen appliances such as electric mixers and light dimmer switches operate by altering the current at the load by increasing or decreasing the resistance of the circuit. Pushing the various buttons on an electric mixer can change the mode from mixing to beating by reducing the resistance and allowing more current to be present in the mixer. Similarly, turning a dial on a dimmer switch can increase the resistance of its built-in resistor and thus reduce the current.

Ohm's law in conclusion is a great tool not only in creating useful thing but also in saving life by determining the cause and effect   of using electricity.

References

Silverio, A.A., Bernas G.D. (2012) Physics: *Exploring Life Through Science* (2nd ed.) Philippines: Phoenix Publishing House. ISBN 978-71-06-3092-9

Salmorin, L.M. (2004) Science and Technology PHYSICS (Updated ed) Philippines: Abiva Publishing House, Inc. ISBN 971-553-154-7

Padua, A. L. , Crisostomo, R. M. (2010) Practical and Explorational Physics (2nd ed.) Philippines: Vibal Publishing House. ISBN 978-971-07-2566-3

Voltage, Current, Resistance, and Ohm's Law. (n.d.). Retrieved February 12, 2016, from https://learn.sparkfun.com/tutorials/voltage-current-resistance-and-ohms-law

Untitled Document. N.p., (n.d.). Retrieved March 19, 2016, from http://www.carbonpower.com/Bill%20Calculation.html

Ohm's Law. (n.d.). Retrieved March 20, 2016, from http://www.physicsclassroom.com/class/circuits/Lesson-3/Ohm-s-Law

PHOTOS CREDITS:
Ohm's logo. Retrieved February 10, 2016 from:
http://powerlisting.wikia.com/wiki/Momentum_Manipulation
Super heroes. Retrieved February 10, 2016 from:
http://www.supermantv.net/messageboard/article1595.htm
Georg Simon Ohm. Retrieved February 12, 2016 from:
http://images.fineartamerica.com/images-medium-large/georg-simon-ohm-german-physicist-sheila-terry.jpg
Ohms Triangle. Retrieved February 12, 2016 from:
http://media1.shmoop.com/images/common-core/Ohms_law_triangle.png
Ammeter. Photo credit to the author
Voltmeter. Photo credit to the author.
Ohmmeter. Retrieved February 12, 2016 from: http://2.bp.blogspot.com/-U5WXWGrFHl4/UUF4lACkiI/AAAAAAAAANc/X8IcG2iY0R4/s1600/gambar+ohmmeter.jpg
Lie Detector Machine. Retrieved February 12, 2016 from:
https://tse1.mm.bing.net/th?&id=OIP.M5d3bdcd2384155ebaa5c3945f2d733d1H0&w=264&h=161&c=0&pid=1.9&rs=0&p=0
Open circuit. Photo credit to the author
Closed circuit. Photo credit to the author
Circuit diagram. Photo credit to the author via phet simulation.
Figure 1 and 2. Photo credit to the author
Power formula. Photo credit to the author
Meralco app. Retrieved March 19,2016 apps.meralco.com.ph/appcal/
Electric meter. Retrieved March 20, 2016 www.fryeelectricinc.com

YOUR KNOWLEDGE HAS VALUE

- We will publish your bachelor's and master's thesis, essays and papers

- Your own eBook and book - sold worldwide in all relevant shops

- Earn money with each sale

Upload your text at www.GRIN.com and publish for free

YOUR KNOWLEDGE HAS VALUE

- We will publish your bachelor's and
  master's thesis, essays and papers

- Your own eBook and book -
  sold worldwide in all relevant shops

- Earn money with each sale

Upload your text at www.GRIN.com
and publish for free